negweny@hotmail.com

Picture of the Grand Duc of Luxembourg's family

Source: ww.fiabci65.com

Picture of His Highness Prince Guillaume of Luxembourg and his wife HRH Princess Stephanie

Source: media1.popsugar-assets.com

Picture of Luxembourg's Prime Minister Mr. Xavier Bettel

Source: www.aldeparty.eu

Picture of Luxembourg's Minister of Health Mme. Lydia Mutsch

Source: images.newmedia.lu

Picture of Luxembourg's Minister of Environment, Mme Carole Dieschbourg

Source: www.paperjam.lu

Map of Luxembourg with the three colored flag

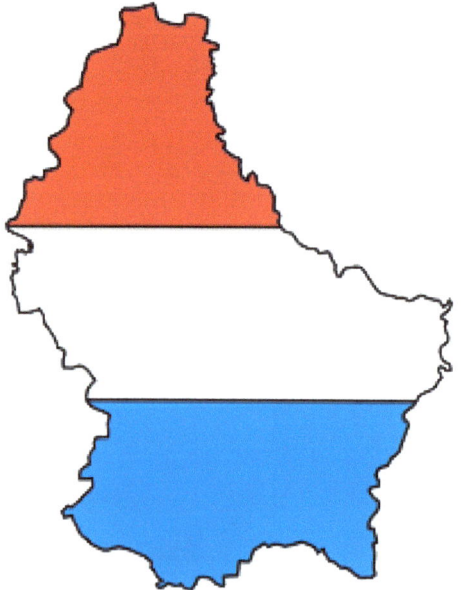

Information about Luxembourg:
Surface area: 2,586 km²
Population (2013): 543,202
Gross domestic product (2013): 60.13 billion US dollars

Source: mapsof.net/map/luxembourg-flag-map

Map of the European Union

Source: commons.wikimedia.org

Information about the European Union:
Surface area: 4 million km²
Population: 503 million
Gross domestic product (2013): €13.070 trillion

Picture of vineyards overlooking the river Moselle in Luxembourg

Source: www.wine-pages.com

Pictures of the esplanade in Remich, Luxembourg, overlooking the river Moselle. The swans, the ducks and their kids are enjoying the esplanade and the Moselle.

Source: www.wort.lu   **Source**: www.panoramio.com

Source: nosamisaplumes.xooit.be

Picture of Remerchen's recreational center (Luxembourg's beach). Please note the cleanliness, restaurant, etc. There are no cigarette buds/megots on the sand.

Source: http://www.wunnen-mag.lu/

Preface

Dear Reader,

Thank you for reading my book 'Not On The Floor'.

Story of the book

In fact, I never ever thought about writing this book or any other book. I am nor a writer, nor an artist... I am a pharmacist and a stamp collector.

The story of this book started in 2009 and was completed in 2015. I was given the unique opportunity to collect 60 000 cigarette buds ('megots' in French) from the floor...

Thank you for reading my book
'Not On The Floor'
because cigarette buds on the floor
=
49€

Pictures of people smoking in the park, at the beach, in the balcony, in the car, on the train, on the beach, etc.

Source: www.smokersworld.info

Source: www.bbc.com

Source: www.telegraph.co.uk

Source: www.vintagetravel.co.uk

In 1500 days, I collected a total of ± 60 000 cigarette buds/megots. This means that on average 40 cigarette buds were collected per day. At first, I was simply enjoying picking up every single bud and keeping it in an empty plastic water bottle of 1.5 litres. With time I had collected 10 plastic bottles, each containing 750 cigarette buds/megots or a total of 7500. I found myself stuck with these 7500 cigarette buds, not knowing what to do with them.

I was trying to figure out the best solution to get rid of these cigarette buds. That's when I contacted Lamesch, the main recycling station in Luxembourg, in order to ask them how to best get rid of these cigarette buds. Lamesch told me that in fact, there is currently no system to collect or recycle cigarette buds/megots.

Of course I asked Lamesch what they do with cigarette buds? Lamesch replied that cigarette buds are already part of the unrecyclable materials, which all get burned.

I asked myself why am I collecting and separating these cigarette buds, when in the end they get burned with all the other unrecyclable materials...

I called my daughter Diana in order to do a Google search to find out the best way to recycle cigarette buds/megots. Diana surprised me after her Google search, because I am not the only one who collects cigarette buds/megots.

Indeed, there is an American environmental activist in San Francisco, who pays 3$ per 1000 cigarette buds/megots, without having a solution on how to get rid of them. However, he never burns them, because cigarette buds are made from plastic.

In fact, I was shocked that filters were made from PLASTIC and I remembered the song that my children used to love so much, which goes something like:
Life in plastic,...it's fantastic!
But for me, I have to sing:
Filter is plastic,...it's not fantastic!

So the first primitive artwork I made says:
> Life in plastic, it's fantastic
> This artwork was made from ±150 handmade rolled up cigarette buds, without filter.

The second artwork says:
> ❖ Filter is plastic, not fantastic.
> This artwork was made from ±150 cigarette buds/megots.

Picture 1

Dimension: 80 x 20 cm
Subject: Life in plastic, it's fantastic.
Quantity: ± 150 handmade rolled up cigarette buds, without filter
Artwork: Negweny
Year: 2014

Picture 2

Dimension: 80 x 20 cm
Subject: Filter is plastic, not fantastic.
Quantity: ± 150 cigarette buds/megots with filter
Artwork: Negweny
Year: 2014

<u>Mr Andre Schramer</u>

I showed my two artworks to Mr Schramer and I informed him about the San Francisco environmental activist and his collection of trillions of cigarette buds/megots, unrecyclable material, without solution.

So Mr Schramer told me: But at least he eliminated them from the environment and one day, there will be a solution.

I replied to Mr Schramer: What about me?

He smiles and said what about?

I replied: Cigarette buds/megots. My collection of at least 7500 buds, what can I do with them?

Mr Schramer explained to me that one of the best messages to people is via ARTS. It gives a far greater effect than dictating what is right or wrong. That's when Mr Schramer bought me a giant wooden frame of ±1.2 m x 1.2m, with the design of a skull with two bones drawn on it. He advised me to stick my collected cigarette buds/megots on it in order to form an artwork.

I followed Mr Schramer's advice and finally I was able to use up all 7500 cigarette buds/megots in this artwork. This was the first artwork made from such a huge amount of cigarette buds/megots.

Picture 3

Dimension: 1.2 x 1.2 m
Subject: Skull with two bones
Quantity: ± 7500 cigarette buds/megots
Artwork: Negweny
Year: 2014

For Mr Schramer and I, and many more who saw this artwork, consider it as a piece of art or a masterpiece, because of the fact that it was purely made from cigarette buds/megots with and without filters (hand rolled). Please note the color shades from bright orange to dark orange in cigarette buds/megots with filter, and bright white to brownish white in the hand rolled cigarette buds/megots. These color changes are due to the moisture of rainwater.

The relation with Mr Schramer continued from 2009 until now, supporting me with all the necessary materials, designs, and best of all, encouragement to proceed with further study in the field of environmental protection, with the aim to issue this book. The principal message of this book is to highlight one fact only:

Not On The Floor
w.e.g., please, s.v.p., Bitte, Por Favor

Dear Reader, I hope you are enjoying reading through this book and are spreading the action of:

Not On The Floor

Purpose of my book

In fact, the main purpose of issuing this book was not to promote an anti-tobacco action, nor a non-smoking habit, or any other attempt to accuse smokers in any way.

The main purpose of this book is just to fill knowledge gaps, which were missing in relation to the habit of smoking.

<u>Missing information</u>:

These mainly relate to information and facts dealing with the habit of smoking, as well as to environmental issues.

There is no reason to accuse or blame smokers for any negative effects on the environment in relation to the habit of smoking.

The only ones to blame are tobacco manufacturers. It is their pure responsibility to broadcast the negative environmental effects, emerging from smoking their products.

Tobacco manufacturers are the only ones who have complete information about their products.

Picture 4

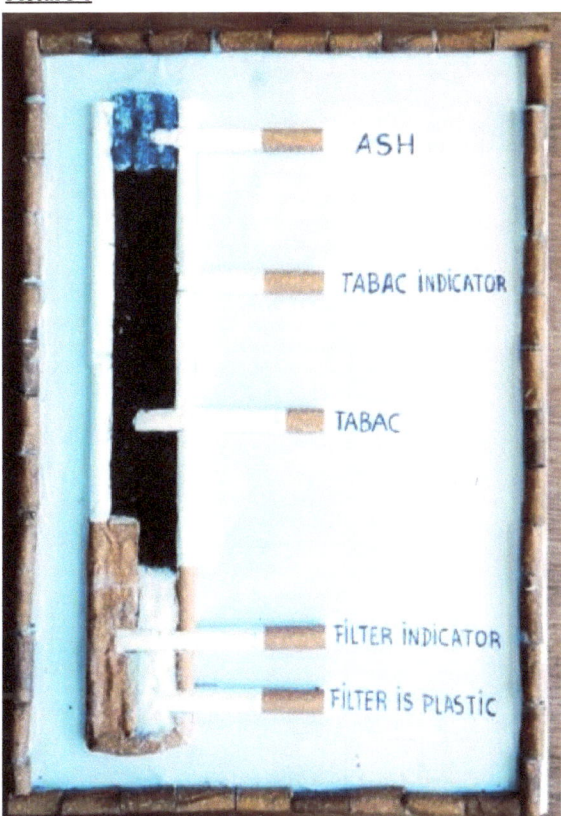

Subject: Cigarette diagram

Picture 5

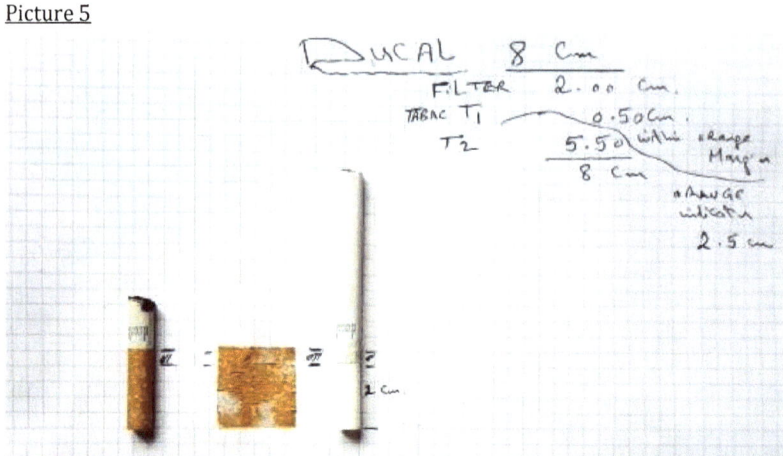

Subject: Ducal with explanation on each length

Picture 6

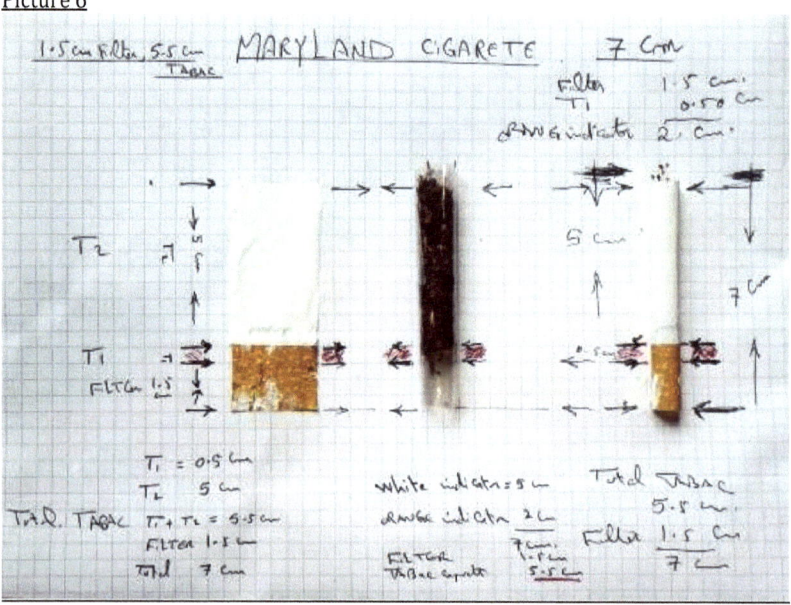

Subject: Maryland with explanation on dimensions

Picture 7

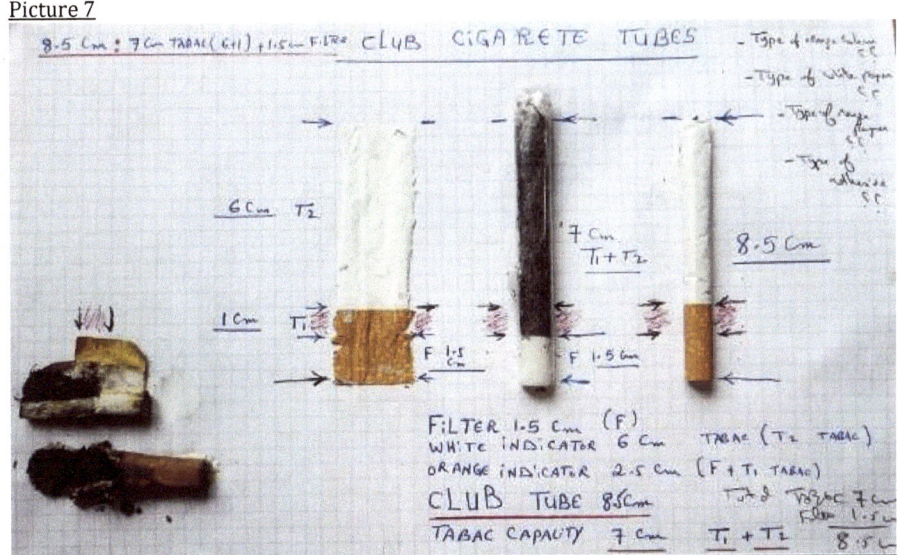

Subject: Club with explanation on lengths

Picture 8

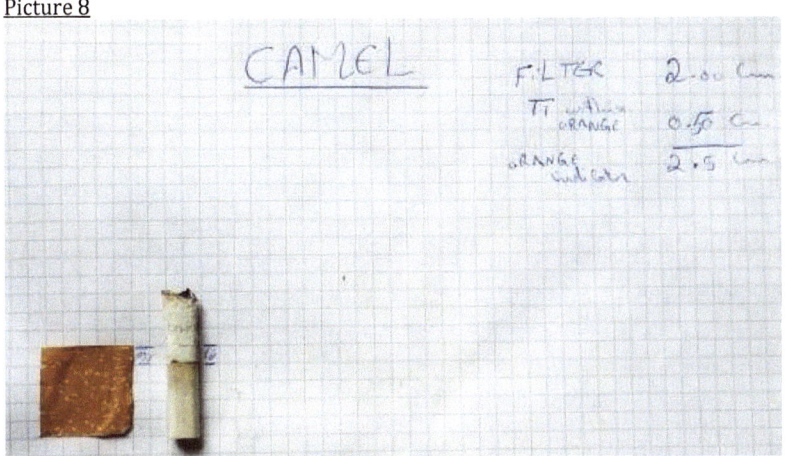

Subject: Camel with explanation on lengths

Dimension: 80 x 80 cm
Design: Mr Andre Schramer
Subject: Picture of Camel photographed by Adolf at home.
Quantity: ± 3500 cigarette buds/megots
Artwork: Negweny
Photographer: Adolf

Picture 9

Subject: Luckies with explanation on lengths

Picture 10

Subject: Marlboro with explanation on length

Picture 11

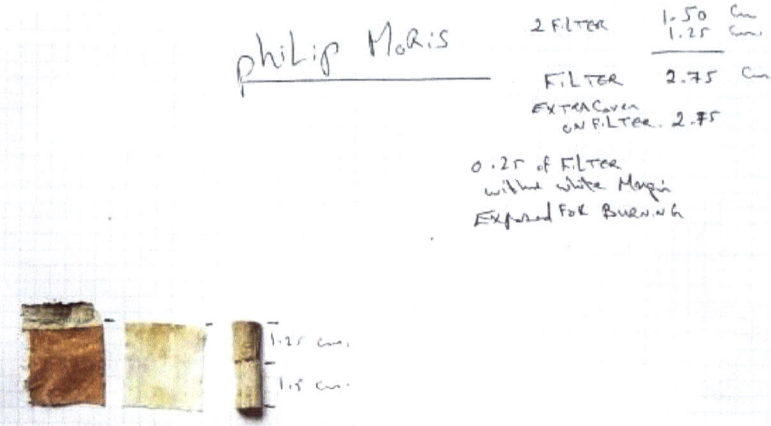

Subject: <u>Philip Morris with explanation on lengths</u>

Picture 12

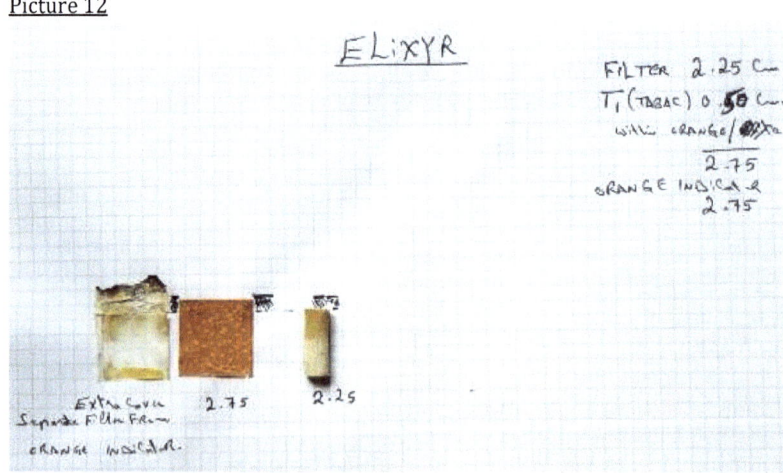

Subject: <u>Elixyr with explanation on lengths</u>

Smoking

It is a habit. Humans smoke, animals smoke too. By the way, a long time ago, I was visiting a zoo with my family in Amneville, and there was a chimpanzee, who was given a cigarette by a visitor, so the chimpanzee was smoking the cigarette while looking at the visitors. After finishing his smoke, the chimpanzee was very angry and started to hit his head, because he didn't know where to put his cigarette bud/megot.

Dimension: 80 x 80 cm
Design: Mr Andre Schramer
Subject: Picture of Chimpanzee (featured on the front cover)
Quantity: ± 4500 cigarette buds/megots
Artwork: Negweny
Year: 2015
Photographer: Adolf

Habits

It is an action we love to do or we are just used to doing it, and sometimes we need to do it.
Eating is a habit. Eating sweets is a lovely habit and no harm of doing it, except if we suffer from side effects, such as obesity, diabetes, etc.
That's why manufacturers replace sugar by artificial sweeteners.
Nowadays, food manufacturers indicate the exact kilocalories (kcal) contained in their products. Not only that! They also make consumers aware of any allergens that might be accidentally contained in the foods, such as traces of nuts.
Food manufacturers even indicate how to dispose or recycle food containers, wrapping materials, etc.
SMOKING is also a habit, but when comparing food and cigarette manufacturers, it can be noted that food manufacturers fulfill their responsibilities towards their consumers. However and unfortunately, cigarette manufacturers bear zero level of responsibility towards their consumers, with the exception of their strong warning that cigarette smoking can kill.

Responsibility of cigarette/tobacco manufacturers

Cigarette and tobacco products can be classified as foods or medicines without prescription, because they contain active constituents, such as nicotine, producing pharmacological or therapeutic effects.

Note to reader: Please refer to any search engine, such as Google or Yahoo to find out more about the pharmacological/therapeutic effects of nicotine.

So because of the above reason, the consumer/smoker has got the pure right to be informed about what he/she is smoking.
It is not enough to state the percentage of nicotine level in tobacco products. Tobacco and cigarette manufacturers should be obliged to indicate the full analytical data of each batch number.

Examples could include:
- Percentage of nicotine, side effects, contraindications to the use of nicotine, and drug/drug interactions between nicotine and other drugs, such as cardiac or respiratory medications.
- Percentage of other chemical constituents and whether they are compatible with the limits allowed for human use.
- Analysis of the ash produced with a way to dispose of it.
- Type of filter material with a way to dispose of it.
- Full diagram of a cigarette, indicating the weight of tobacco and the size of the filter.

Cigarette remains:

Cigarette remains are called:
1. cigarette buds (English)
2. megots (French)
3. Stomp (Luxembourgish)

Composition of cigarette remains:
In fact, environmental activists have long been dedicating their time mainly on cigarette buds/megots.
However, as we know, cigarette remains consist of 3 main elements, plus of course the smoke, which is inhaled by the smoker.
The 3 main elements of cigarette remains are:
1. ashes
2. remains of unsmoked tobacco, found within the orange part of the cigarette (the filter indicator)
3. the filter itself

The main remains of cigarettes:

1. ASHES

Ashes are the end product, after complete burning of cigarette contents. I like to be accurate when I say complete burning of cigarette contents, rather than saying complete burning of tobacco.

Why? As we all know, cigarette contents do not only contain ground tobacco leaves, but also other additives, including chemical and organic products.

Dear Reader: Kindly use any search engine, such as Google or Yahoo to identify the different types of cigarette additives.

Every cigarette or tobacco manufacturer has its own formula of the contents of their cigarettes or tobacco. That's why it is the consumer's/smoker's pure right to know exactly what he/she is smoking.

Hence, the ashes produced as a result of burning cigarettes/tobacco will differ from brand to brand. Manufacturers should be obliged to inform the consumer/smoker about the full chemical analysis of the produced ashes.

Picture 13

Subject: <u>Ashes</u>
Quantity: 102 grams
Number of cigarettes burned: 600 cigarettes
Percentage: ± 15%
Date: collected Nov. 2014

Picture 14

Subject: Ashes, same as in picture 13, but included in an empty glass bottle of Nescafe of 200g. Completely full at time of collection.
Quantity: 102 grams
Number of cigarettes burned: 600 cigarettes
Percentage: ± 15%
Date: collected Nov. 2014

General physical character of Ash:

Ashes are very fine, greyish white, fluffy particles. When freshly produced, they consist of microparticles flying unnoticed in the air. In my opinion, they can diffuse through our lungs. When shaking the ash within the glass bottle, it produces a cloud, which is easily inhaled.

Ashes have an offensive smell when collected in quantity. I am not sure whether ash can diffuse into our blood circulation.

However, what I am 100% sure about, is that if ash contains any water soluble contents, that the exchange will happen at our lung walls.

That's why it is so important that cigarette/tobacco manufacturers state the complete chemical analyses of their specific ashes. Otherwise, we remain blinded to the health risks the entire population is exposed to, whether smokers and non-smokers.

As previously mentioned, ashes are fluffy when freshly produced, but become more and more compacted to almost 50% of their original volume within 6 months. Please have a look at the next 4 pictures and compare them with their original volume in Picture 14.

Picture 15

Subject: <u>Ashes</u>, same as in picture 14.
Its volume was reduced 2 months later.
Quantity: 102 grams
Number of cigarettes burned: 600 cigarettes
Percentage: ± 15%
Date: collected Jan. 2015

Picture 16

Subject: <u>Ashes</u>, same as in picture 15. Its volume was reduced 2 months later.
Quantity: 102 grams
Number of cigarettes burned: 600 cigarettes
Percentage: ± 15%
Date: collected March 2015

Picture 17

Subject: Ashes, same as in picture 16. Its volume was reduced 2 months later.
Quantity: 102 grams
Number of cigarettes burned: 600 cigarettes
Percentage: ± 15%
Date: collected May. 2015

Picture 18

Subject: <u>Ashes</u>
Quantity: 96 grams
Number of cigarettes burned: 700 cigarettes
Percentage: ± 14%
Date: collected Oct. 2014: photo was taken in June 2015

At the time of collection in Oct. 2014, the volume of the ash was filling the whole glass bottle, before being sealed. However, when taking the photo in June 2015, the ash was almost reduced to 50% of its original volume and became compacted within 8 months.

The weight of ash:

Let's assume that the weight of a cigarette is 1gram. This means that ±0.2grams of ash is produced when a cigarette is completely smoked up to the filter indicator limit (the full white part up to the start of the orange margin).

In Luxembourg, the total annual sale of cigarette/tobacco is 500 million Euros. Out of these, 150 million Euros are from tobacco products used locally and the remaining 350 million Euros are from tobacco products consumed by commuters from the 3 frontiers: Belgium, France, and Germany. This means that Luxembourgish smokers spend 150 million Euros per year, with the average price of a cigarette packet being ±5€. Hence, the total number of packets sold annually in Luxembourg is ± 30 million. Each packet contains at least 20 cigarettes, meaning that smokers in Luxembourg smoke an average of 600 million cigarettes per year (30 million packets x 20 cigarettes).

Let's assume once more that a cigarette weighs 1gram. This would mean that Luxembourg burns ± 600 000 000 grams, equal to 600 tons of tobacco, per year. 20% of a cigarette's weight results in the production of ash, upon complete burning of cigarettes up to the filter margins. So, Luxembourg produces 120 tons of ash per year.
If the European Union represents 1000 times Luxembourg's population, then the EU produces 120 000 tons of ash per year.

Ash in Luxembourg
Let's assume that the population of Luxembourg is ± 500 000. This would mean that 20% are smokers, smoking a total of 600 million cigarettes per year. So each smoker smokes ±6000 cigarettes per year or ±500 cigarettes per month.
Each smoker produces 20% ash per cigarette weight, so a total of 100 grams of ash per month. So we have got a total of 10 tons of ash produced per month or 120 tons of ash produced per year.
Dear Reader,
Each of the 100 000 smokers in Luxembourg produces one full bottle of ash per month (± 100 grams). So we get 100 000 (HUNDRED THOUSAND!) bottles of ash per month. And we get 1 200 000 (ONE MILLION TWO HUNDRED THOUSAND!) bottles of ash per year.
****The EU produces one billion, two hundred million bottles of ash per year (1 200 000 x 1000).****

WHERE DOES ALL OF THIS ASH GO?

Where are the ashes?

As far as there is no way to collect cigarette/tobacco ashes, these ashes are totally free ashes, which are very very light.

Ashes can fly long distances with the winds.

Ashes are composed of end burning products of tobacco and contain additives (unknown materials), traces of pesticides, insecticides, etc.

From my point of view, ashes are the most dangerous phenomenon when we deal with the habit of smoking.

Smoking can have an effect on the smoker, who inhales the smoke, which contains nicotine and other unknown additives. Smoking can also have an effect on other people surrounding the smoker, via passive smoking.

However, ashes can have a direct effect on every single human, animal, bird, fish…, living in our country Luxembourg.

Following the anti-tobacco legislation of no smoking in public places, smokers smoke in the open air.

At home, we smoke in the balcony, by the window, in the garden, etc. In a car, we smoke and let the ashes escape from the window to reach innocent people, who are taking a walk to get some fresh, clean, Luxembourgish air.

A very logical question to ask the Minister of Environment: Where did our 120 tons of ashes produced last year go?

Ash was burned

I agree that we are educated and civilized enough to dispose of our cigarette buds/megots in ashtrays when smoking in cafes, banks, restaurants, hotels, etc. These ash trays are then emptied into rubbish bins and collected by Lamesch trucks, crushed at every stopover.

As previously mentioned, ashes are very light and can easily fly. So ashes will fly and fly and will first attack the truck workers, despite using masks.

Also ashes will attack innocent mothers and their babies. Ashes will attack elderly people during their walks.

Let's suppose ashes arrive safely to a Lamesch site, to be added to other unrecyclable material, ready for burning. Ashes are already burned material, so we would burn it again, probably using very high temperatures… In fact, I don't know what the end product of such a process is… I leave it to scientists to start their efforts to try to protect our Luxembourgish fresh air. Ash may be an important culprit for the increase in the number of asthma patients.

My message to smokers is to kindly dispose their cigarette ashes, Not On The Floor… You might ask: Where then? I reply: I don't know, but Not On The Floor ☺

w.e.g.

The main remains of cigarettes:

2. **Remains of unsmoked tobacco found within the orange filter indicator:**

After a detailed study on my 60 000 collected cigarette buds/megots, I realized that there is a huge amount of unused tobacco. I call it second hand tobacco, because it is not the same quality of manufacture grade tobacco, having suffered from all gases and trapped materials through the smoking process.

The percentage of unused tobacco can reach up to 30% in Luxembourg, because of the lower price of tobacco products and the higher level of public income. That's why, I am sure that the percentage will be much lower in Germany or in Bulgaria.

Picture 19

Subject: Remains of unused tobacco from the orange filter indicators
Quantity: 2.7kg
Number of cigarettes: 15 000
Date: January 2015

Picture 20

Comment: Simple diagram showing different lengths of unsmoked tobacco. This means that smokers have different behaviors: some smoke just a little, others smoke up to the filter indicator; and others smoke even up to the plastic filter itself, which may be a strong risk factor for cancer. Please note the wrong side of burning the cigarette at the lower line. Already smokers are smoking part of its plastic filter ☹.

Picture 21

Comment: Uncovered ashtray at Belval, Luxembourg with cigarette remains with a large amount of unsmoked tobacco. It attracted our friend the grasshopper, who started eating the unsmoked tobacco.

It might be a good idea to cover public ashtrays, as well as private ones, in order to stop ash from circulating into our fresh Luxembourgish air and to stop grasshoppers and others from reproducing and attacking our agricultural products.

Why unused tobacco?

1. Cigarette manufacturers' precautionary measure:

Actually, it is with a good intention that cigarette manufacturers are trying to safeguard smokers' risk of smoking the filter, which is made of plastic.

So as a precautionary measure, almost all cigarette manufacturers include a certain length of tobacco within the orange filter indicator.

The length of tobacco within the orange filter indicator varies:

- ❖ As per my pictures 5, 6, 8, 10, and 12: Ducal, Maryland, Camel, Marlboro, and Elixyr leave 0.5cm of tobacco within the orange filter indicator.
- ❖ As per my picture 7: Club leaves 1cm of tobacco within the orange filter indicator.
- ❖ As per my pictures 9 and 11, Luckies and Philip Moris leave no tobacco within the white filter indicator (Luckies) and within the orange filter indicator (Philip Moris). Moreover, they include 0.25cm of filter length within the tobacco length indicator. This means that smokers smoke plastic for a length of 0.25cm, as per these two pictures.

Picture 5

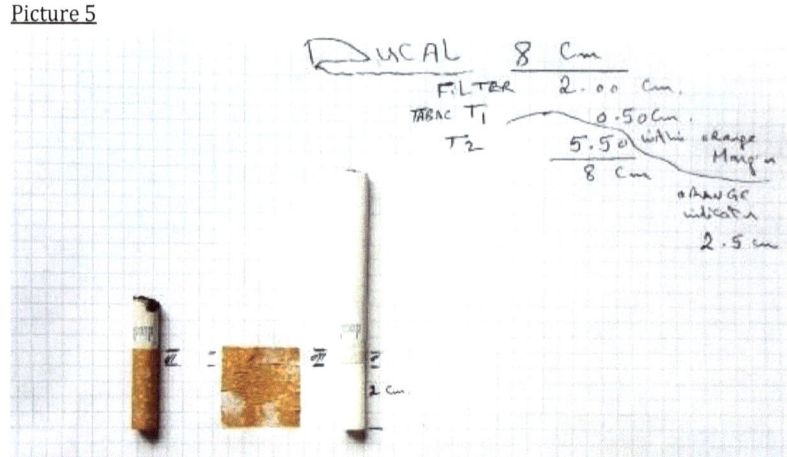

Comment: Ducal Diagram
Precaution: Unsmoked tobacco within the orange filter indicator: 0.5cm.

Picture 6

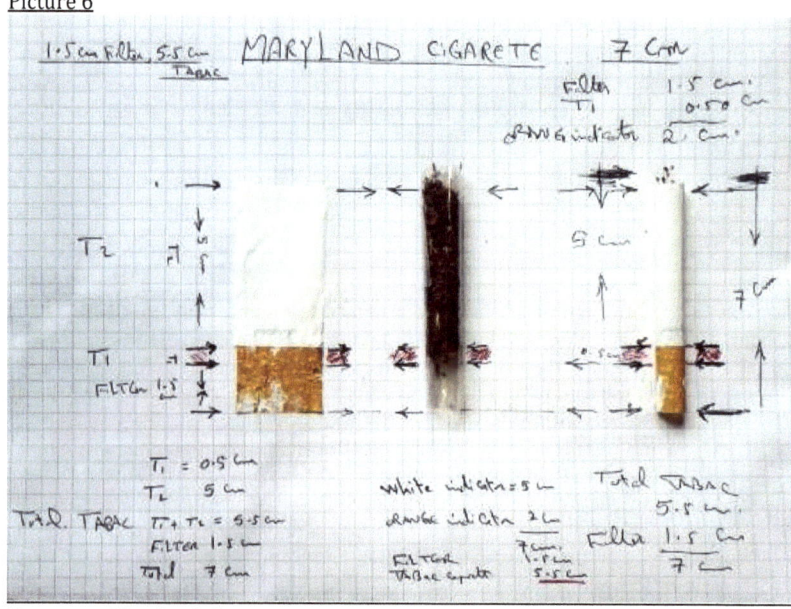

Comment: Maryland diagram
Precaution: Unsmoked tobacco within the orange filter indicator: 0.5cm.

Picture 8

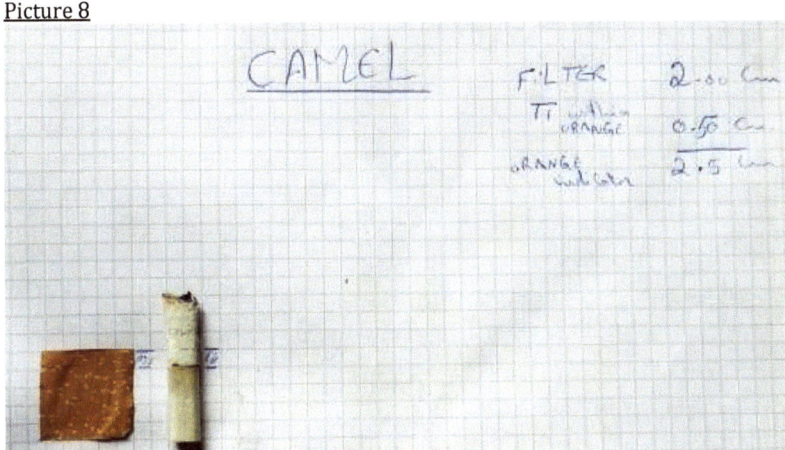

Comment: Camel Diagram
Precaution: Unsmoked tobacco within the orange filter indicator: 0.5cm.

Picture 10

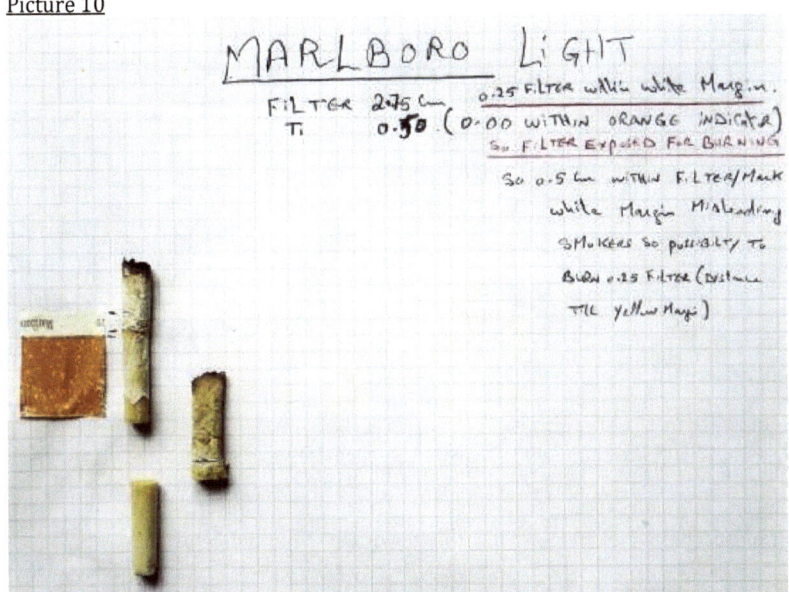

Comment: Marlboro Diagram
Precaution: Unsmoked tobacco within the orange filter indicator: 0.5cm.

Picture 12

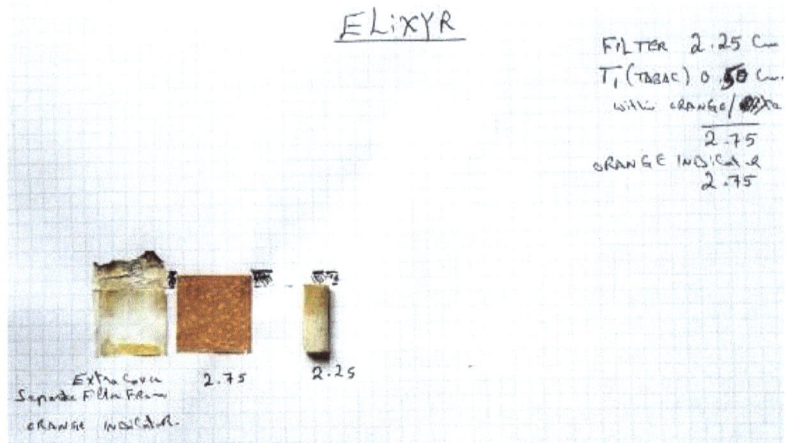

Comment: Elixyr Diagram
Precaution: Unsmoked tobacco within the orange filter indicator: 0.5cm.

Picture 7

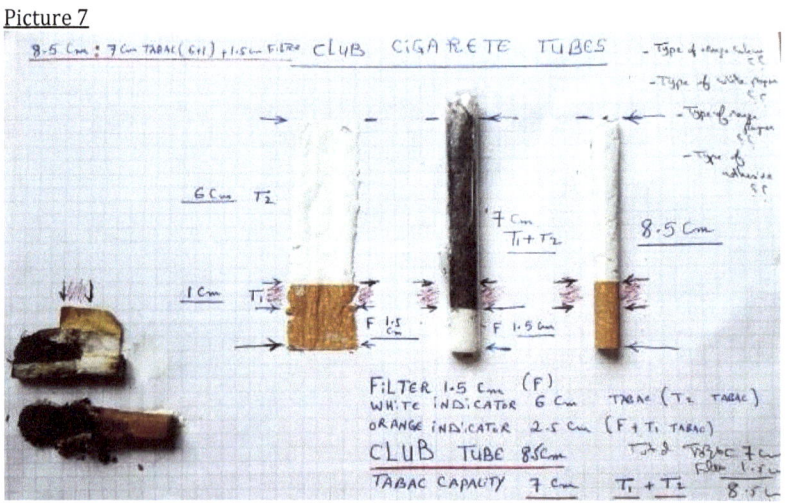

Comment: Club Diagram
Precaution: Unsmoked tobacco within the orange filter indicator: 1cm.

Picture 9

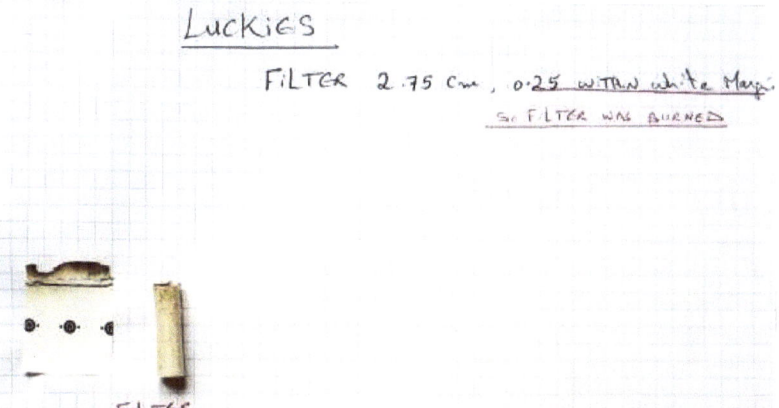

Comment: Luckies Diagram
Precaution: Unsmoked tobacco is ZERO. Even 0.25cm of filter length is included within the white margin filter indicator exposing smokers to the risk of smoking 0.25cm of plastic filter.

Picture 11

Comment: Philip Moris Diagram
Precaution: Unsmoked tobacco is ZERO. Even 0.25cm of filter length is included within the white tobacco margin indicator, exposing smokers to the risk of smoking 0.25cm of the plastic filter.

Why unused tobacco?

2. Smokers hurry to smoke:

Following the recent legislation prohibiting smoking in public places, offices, cafes, restaurants, etc., smokers are enforced to smoke outside in the open air. Depending on the location, in most cases, smokers are unable to smoke the entire cigarette length of tobacco, because of the following reasons:
- Lack of time
- Raining in the open air
- Snowing
- Feeling cold in cold weather
- Feeling hot in hot sunny weather
- Finding oneself enforced to smoke in the presence of a person we are trying to avoid
- Being surprised with his/her boss coming down to smoke
- For any other reason, such as a smoker being rich enough to be able to buy more cigarettes, smoking only a few centimeters each time and leaving the remainders for grasshoppers to eat ☺.

*** So smokers are enforced to not complete smoking the whole cigarette length, leaving several centimeters of unsmoked tobacco. ***

Why unused tobacco?

3. Smokers are trying to quit smoking:
Some smokers are willing to quit smoking, whether because of a medical reason or any other reason. That's why they try to stop smoking the entire length of the cigarette and end up smoking 4cm, then 3cm, 2cm, 1cm, until they fully stop smoking. This progressive reduction in the length of cigarette smoked leads to remains of unsmoked tobacco.

Luxembourg and unused tobacco
In fact, it is a problematic, leaving brown stains on the floor. Luxembourg smokes 600 million cigarettes per year, so about 600 tons of tobacco. Unused tobacco makes up about 30%, meaning that we dispose about 180 tons of tobacco a year into the environment...
...or 300g of tobacco per minute
...or 500kg per day ☹.

Picture 22

Comment: Unused tobacco within the orange filter indicator varies in size between 0.5cm for Ducal, Marlboro, Maryland, Camel, and Elixyr, and 1cm for Club cigarettes.

Picture 23

Comment: Unused tobacco within the white tobacco indicator and also within the orange filter indicator: more than 1cm in length, sometimes reaching up to 5cm in length of unused tobacco. Please note that what you see is just unused tobacco, without the filter!

Pictures 24 and 25

Comment: This pink colored statue contains 700gm of unused tobacco, extracted from 5000 cigarette buds/megots. It was a new idea to dispose unused tobacco this way, as artworks continuously broadcast their odor, irritating people in the surroundings ☹.

Picture 26

Comment: What you see is just unsmoked tobacco: pure tobacco found within the orange filter indicator and large part within tobacco white indicators.
Length: 50cm
Height: 40cm
Approximate unused tobacco weight: ± 400g.

Picture 27

Comment: What you see is the sun rising at the beach. It was made from unused tobacco: pure tobacco varies in length from 0.5cm to several centimeters, found within the orange colored filter indicator and within the white tobacco indicator. This artwork was purely made from tobacco, without any filter.

Length: 50cm
Height: 80cm
Approximate unused tobacco weight: ±700g.

Pictures 28 and 29

Comment: Luxembourgish Egyptian pyramid under construction
Inside this pyramid under construction, there is a metal container internally covered with aluminum foil and containing 2.7kg of unsmoked tobacco remains, extracted from 15 000 cigarette buds/megots.

Unused tobacco and its environmental effect

A – Effect on our water resources:

As previously explained, there is a huge amount of unused tobacco in cigarette buds/megots. When those cigarette buds/megots are thrown on the floor, in forests, on beaches, etc., they get exposed to moisture from rain, snow, etc. That's why the white-orange margins are very easily opened, thereby exposing the unused tobacco to the atmosphere. Rain or snow then wash unused tobacco into the water draining systems in the streets, causing serious pollution. Tobacco products include additives, added by cigarette manufacturers. For example, insecticides and pesticides can be found on tobacco leaves to help protect against rodents, insects, etc. during the cultivation and fermentation processes of tobacco leaves. So it is not a surprise to detect slight percentages of pesticides/insecticides in our water resources.

B – Spreading of insects from other countries:

As we know, tobacco leaves are very large with a very wide surface. This means that they are very attractive to insects to live on, to eat from, and to reproduce on. Examples include grasshoppers, ants, spiders, etc.

As we know, tobacco is cultivated in warm countries, such as Africa, Asia, and South and North America. By importing tobacco leaves, eggs of insects, such as ants, spiders, or grasshoppers, can be found on these tobacco leaves.

By exposing unused tobacco to raining/snowing conditions, followed by warm weather, we are helping the deposited eggs to mature and develop foreign insects in host countries, such as Luxembourg.

So it is not strange to see strange insects in Luxembourg during hot Summer days.

Picture 21

Comment: Grasshopper eating tobacco left in a hurry by smokers at Belval, Luxembourg. I don't know from which country this grasshopper comes from. But this grasshopper was just lucky to find its oriental food in Belval.

The main remains of cigarettes:

3. Cigarette filter:

Most of cigarette manufacturers include filters to protect smokers from poisonous materials, such as lead, mercury, arsenic, cadmium, hydrocarbons, tar, etc.

***Cigarette filters are mainly made from cellulose acetate, which is plastic, meaning that it is non recyclable material. ***

Filter size varies from manufacturer to manufacturer:
- Club, Maryland: 1.5cm (Pictures 6 and 7)
- Ducal, Camel: 2cm (Pictures 5 and 8)
- Elixyr: 2.25cm (Picture 12)
- Luckies, Marlboro Light, Philip Moris: 2.75cm (Pictures 9, 10, and 11)

Please note that one cigarette filter is able to pollute 500 liters of drinking water.

Recycling cigarette filters:

Until today, there doesn't exist a way to recycle or to reuse cigarette filters. Cigarette filters are very rich in the most dangerous poisonous materials, such as tar, lead, mercury, arsenic, cadmium, acetone... and other hydrocarbons. The only way to get rid of cigarette filters along with tobacco remains is by burning them. So in Luxembourg, we burn 1.6 million cigarette buds/megots a day, made from plastic, polluting our environment, besides the ashes produced by smoking 1.6 million cigarettes a day.

Picture 30

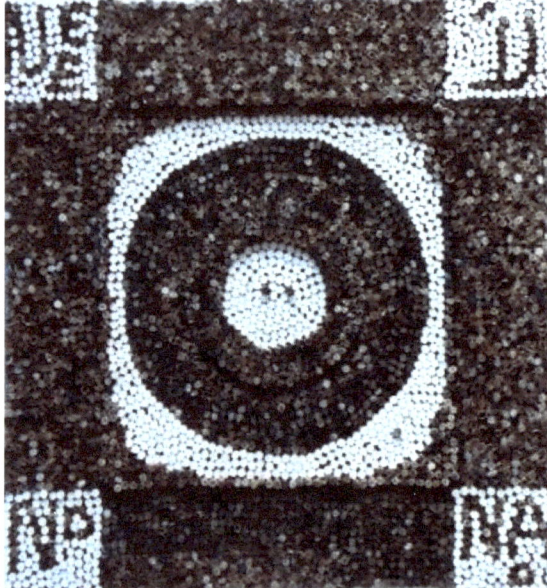

Dimension: 60 x 60 cm
Quantity: ± 5000 cigarette filters
Artwork: Negweny
Arrangement: implanted vertically

Comment: Artwork made mainly from used cigarette filters, which are dark brown in color and include unused Club cigarette filters with bright white color. So the reader can compare between the bright white unused filters and the dirty used ones.

The writing in Luxembourgish: NEE = NO
The writing in Arabic: ﻻ = NO
The writing in Portuguese: NÃO= NO

Picture 31

Dimension: 60 x 60 cm
Subject: NO SMOKING SIGN
Quantity: ± 5000 cigarette filters
Artwork: Negweny
Arrangement: vertically for all filters (used and unused); horizontally for the sign of "no smoking" and for the central cigarette buds/megots, hand rolled without filters.

Comment: This artwork was mainly made from used cigarette filters, using unused Club tube filters. Used filters are dark brown. Unused filters are bright white. We used the unused white filters to write:

"FILTER IS PLASTIC NOT FANTASTIC"
NEE = NO in Luxembourgish
لا = NO in Arabic

So the reader can have a comparison between the unused bright white filters and the dirty used ones. The circle was entirely made from cigarette buds/megots (filters and tobacco remains).

Picture 32

Comment: Artwork made mainly from used cigarette filters, with the outside orange filter indicator removed. This is to expose the color of the filters after their use, which is dark brown to light brown, depending on the length of tobacco used.

The writing was done by using Club unused bright white filters.

Porque ("why" in Portuguese) On The Floor.

NEE = NO in Luxembourgish (made with used filters)

Dimension: 60 x 40 cm
Quantity: ± 1000 filters, with the outside orange filter indicator removed
Artwork: Negweny
Arrangement: Horizontally for used filters; Vertically for the writing, using unused Club filters.

Picture 33

Dimension: 30 x 40 cm
Subject: <u>Megot on the floor = 49€</u>
Quantity: ± 1200 filters
Artwork: Negweny
Arrangement: vertically

Comment: Artwork made mainly from used cigarette filters, using unused Club tube filters. Used filters look dark brown. Unused filters are bright white.
We used the unused white filters to write: "Megot on the floor = 49€", meaing that cigarette buds/megots on the floor get 49Euros penalty (or 68€ in Paris). So the reader can have a comparison between the unused bright white filter and the dirty used filters to become aware that putting one cigarette bud/megot on the floor in Luxembourg will cost you 49€ in cash ☹.

CONCLUSION

Thank you for reading my book, now you have got an idea of the dimension of the problem caused to our lovely Luxembourgish environment, as a result from the habit of smoking.

***IN FACT, UNLESS WE THINK OF A SOLUTION, THIS PROBLEM WILL CONTINUE. ***

My point of view:
There are three main operations NEEDED:
1. collect them
2. block them
3. recycle them

So collecting and blocking cigarette buds/megots is easy. However, recycling them is not yet possible. In order to finance these 3 operations, I propose to the Luxembourgish Government to impose 30 cents on each packet sold in Luxembourg. Luxembourg sells ±500 million euros of cigarettes a year (150 million Euros: local use, 350 million Euros: 3-frontiers use). This means, almost 100 million packets. Imposing 30 cents per cigarette packet would produce 30 million Euros per year for the purpose of recycling.

In the following pages, I will discuss these 3 operations that are needed to dispose of cigarette buds/megots.

1. Collect Them:

The Government or each municipality should call for an enterprise to be responsible to collect
- A – Ashes
- B – Cigarette buds/Megots

Sources:
1. Residents should be able to deliver ashes and cigarettes buds/megots to their municipality's recycling centers.
2. Specialized companies should be responsible for the collection of cigarette buds/megots and ashes from cafes, restaurants, parks, hospitals, schools, shopping centers, Government buildings, etc.

The collected materials should then be delivered to specialized national centers for cigarette buds/megots research.

2. Block Them

A – Ashes

To be kept in a tightly closed container, waiting for scientific study to determine a suitable and safe way to discard them.

B – Tobacco Remains

Luxembourgish smokers produce at least 180 tons of unused tobacco remains. These should be returned to the manufacturers, as they are the only organizations able to find out a way to recycle them, which is also their pure responsibility.

C – Used Cigarette Filters

Crush them to a fine powder to reduce their volume to a minimum. Stock them in an entirely sealed container, awaiting scientific study to determine a safe way to get rid of them.

3. Recycle Them

Luxembourg needs to cooperate with Luxembourgish and international universities and research centers or event to establish:

A National Research Centre Of Cigarette Buds/Megots

At that Center, the necessary studies would be run to find a safe way to get rid of ashes, unused tobacco, and cigarette filters.

Picture of Chimpanzee

WELCOME TO NEGWENY'S ZOO

Represents our environment, saying Not On The Floor

Picture 35: DONKEY

Comment: Our environment is like donkeys. Donkeys accept everything, but at some point they can become angry.

The hair covering the donkey's body came from Negweny's head hair.

The donkey is eating/smoking cigarettes, excreting nicotine and other harmful materials in a urine collector and excreting filters in a container, while saying "Please, w.e.g. recycle it". We reply to him "Jo Jo", meaning "Yes, Yes" in Luxembourgish.

Dimension: 70 x 70 cm
Quantity: ± 1500 used filters and unused Club filters in bright white color
Design: Mr Andre Schramer
Artwork: Negweny

Picture 36: AMSEL

Comment: Our environment is like Amsel. Surrounded by used smelly cigarette filters, it is claiming that his feathers fell down, without a reason. However, Negweny collected the Amsel's feathers over 1500 days and reassembled the Amsel. So the lovely black color is:
REAL AMSEL FEATHERS

Dimension: 30 x 40 cm
Quantity: ± 1100 used filters
Design: Mr Andre Schramer
Artwork: Negweny

Picture 37: BIRDS GET FREEDOM FROM CIGARETTE FILTERS

Comment: Our environment is like birds. They are trying to get rid of the offensive smell from the used plastic filters. The bird itself was made from hand rolled cigarettes, without filters, because the bird doesn't like the smell of filters.

Dimension: 30 x 40 cm
Quantity: ± 900 used cigarette filters and 150 handmade rolled cigarettes without filters.
Artwork: Negweny

Picture 38: ANGRY RAT

Comment: Our environment is like rats. The rat actually lives all its life depending on itself for the supply of food and never expects that we supply it with food. In the contrary, we supply it with a poison to thin its blood. That's why the rat became angry from the offensive smell coming from cigarette filters. The color covering its body is made from a mixture of ashes and toothpaste in order to make the rat happy, despite still looking angry, as you can see in the picture.

Dimension: 40 x 30 cm
Quantity: ± 900 used filters
Design: Mr Andre Schramer
Artwork: Negweny

Picture 39: THE RAT IS STILL ANGRY

Comment: The same angry rat is not happy with the smell of ashes, which is covering its body, as it has an offensive smell. So Negweny in this artwork offered the rat his real head hair, so that the rat becomes satisfied, although still not happy being surrounded by used plastic filters. He needs cheese, not cigarette buds/megots.

Dimension: 40 x 30 cm
Quantity: ± 900 used filters and real head hair from Negweny
Design: Mr Andre Schramer
Artwork: Negweny

Picture 40: FISH

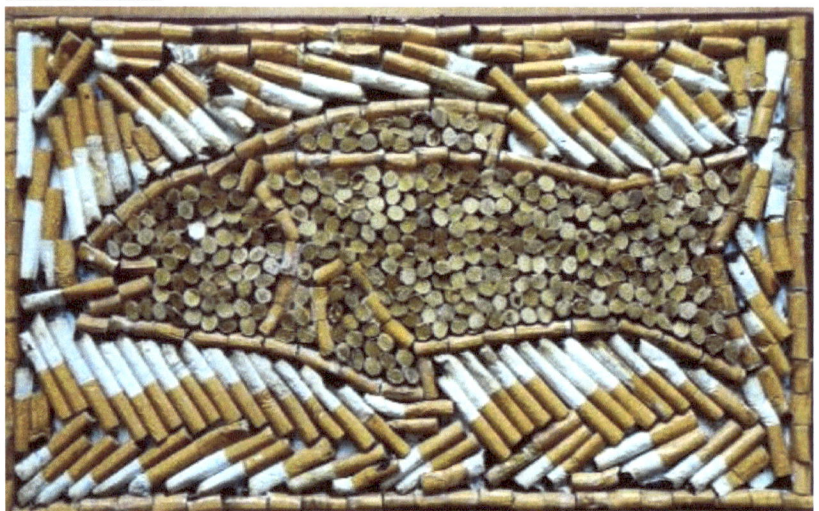

Comment: Our environment is like fish. The fish lives and swims all its life peacefully, but suddenly it found itself surrounded by cigarette buds/megots. Tourists who are rich enough to smoke expensive cigarettes and smoke only one or two centimeters throw their cigarette buds/megots on the beach, which gets washed up to sea. The fish is asking itself whether the hotels accommodating such rich tourists (1 billion tourists per year, worldwide) could supply the tourists with portable ashtrays to prevent the cigarette buds/megots to be thrown on the beach!!!

Dimension: 30 x 20 cm
Quantity: ± 150 cigarette buds/megots ± 200 cigarette filters mounted vertically
Design: Mr Andre Schramer
Artwork: Negweny

CAMEL

Dimension: 60 x 80 cm
Quantity: ± 3500 cigarette filters
Design: Mr Andre Schramer
Artwork: Negweny
Photo: Adolf

Comment: Our environment is like camels. The camel is not happy with such huge amounts of cigarette filters and their offensive smell. The camel has never experienced such smells in its previous generations.

FAMILY OF 3 COWS

Comment: Our environment is like this family of 3 cows: the cow mother and her son accidentally ate the cigarettes, so they became so angry! So the cow father felt it and he went to a place where there are no cigarettes, but only herbs and he was happy saying thank you for megots Not On The Floor.

Dimension: 80 x 60 cm
Quantity: ±4000 cigarette filters
Design: Mr Andre Schramer
Artwork: Negweny
Photo: Adolf

ANGRY ELEPHANT

Comment: Our environment is like elephants. The elephant is so angry with such a huge amount of cigarette filters, which have a totally unacceptable smell.

Dimension: 100 x 80 cm
Quantity: ± 4500 cigarette filters
Design: Mr Andre Schramer
Artwork: Negweny
Photo: Adolf

CHIMPANZEE

Comment: Our environment is like chimpanzees. This is the link between humans and animals smoking cigarettes made in Luxembourg. In reality, the chimpanzee is not angry, because I covered his body with Negweny real head hair, but he requested us to say in each language, including Japanese: "Not On The Floor", so what he really means is: Ok we smoke but cigarette buds/megots should Not go On The Floor. He said it in Japanese: maybe if we can read, it says Not On The Floor.

Dimension: 100 x 80 cm
Quantity: ± 4500 cigarette filters
Design: Mr Andre Schramer
Writing in Japanese: a mixture of ash and toothpaste
Artwork: Negweny
Photo: Adolf

Picture 41

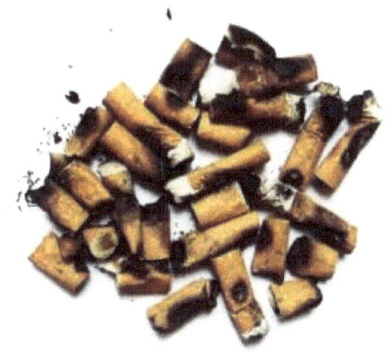

NO COMMENT

Picture 42

NO COMMENT

Picture 43

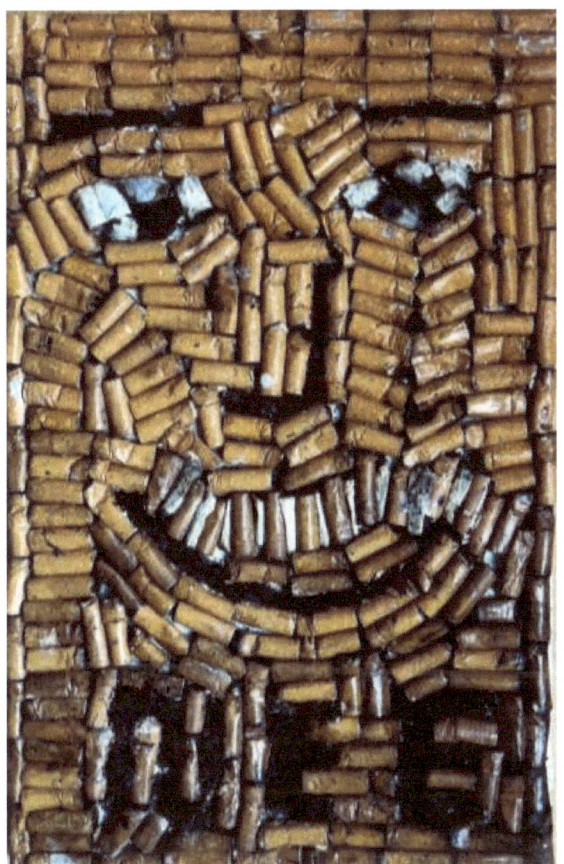

NO COMMENT

Picture 44

NO COMMENT

Picture 45

NO COMMENT

Picture 46

NO COMMENT

Picture 47

NO COMMENT

Picture 48

Cigarette filters arranged vertically

Picture 49

25 x 25 cigarette buds/megots

12 x 12 cigarette buds/megots

NO COMMENT

Picture 50

NO COMMENT

Picture 51

NO COMMENT
This boat statue contains inside its sealed body 500 cigarette buds/megots.

Picture 52

NO COMMENT

This boat statue contains inside its sealed body 500 cigarette buds/megots.

Picture 53

NO COMMENT

This boat statue contains inside its sealed body 500 cigarette buds/megots.

Picture 54

In the 1930s, France was promoting cigarette smoking on postal stamps! Cuba was also promoting smoking cigars on postal stamps! Nowadays, it is no longer permitted to advertise any type of smoking.

You can smoke as much as you like, but most importantly your cigarette buds/megots should be discarded Not On The Floor.

DID YOU KNOW?

Luxembourg	European Union
Smokes 1000 cigarettes per minute	Smokes 1 million cigarettes per minute
Smokes 600 million cigarettes per year	Smokes 600 billion cigarettes per year
Spends 150 million Euros on cigarettes per year	Spends 150 billion Euros per year on smoking
Has 110 000 smokers	Has 110 million smokers
Smokes 60 000 cigarettes per hour	Smokes 60 000 cigarettes per 4 seconds

So what Negweny collected (60 000 cigarette buds/megots) over 1500 days, represents ONE HOUR of smoking time in Luxembourg and represents 4 seconds of smoking time of the European Union.

So please, W.E.G., Bitte, Por Favor
Not On The Floor:
NO ASH
NO TOBACCO
NO FILTERS

Back Cover

It was Not Negweny.
It was them! So a great thanks to ALL OF THEM.

Mr Andre Schramer

Mr Schramer is the owner of the idea of using cigarette buds/megots in creating these artworks, so that the public can view these artworks and receive a lovely message of Not On The Floor. No Ash, No Tobacco Remains, and No Filters On The Floor. Mr Schramer is the designer of the subjects of all these artworks.

Mr Zlatnik Matyas

Mr Zlatnik Matyas is the photographer of all these cigarette buds/megots artworks in this book. His great, excellent efforts are highly appreciated to allow this book to see the light.

My daughter D.C.

She is the Internet researcher gathering all information, related to Ash, Tobacco remains, and Filters. Also, she is the main active action in processing all these pictures and issuing the trial books 1 and 2 from CD1 and CD2.

Smokers

It is through them that more than 60 000 cigarette buds/megots could be collected from the floor. This collection lead to Mr Schramer's inspiration to create the message of the subject of these artworks: "Not On The Floor". So a great Thank You to all smokers.

www.ingramcontent.com/pod-product-compliance
Lightning Source LLC
Chambersburg PA
CBHW040810200526
45159CB00022B/133